SERENGETI PLAIN

Terri Willis

Technical Consultant and Primary Photographer
John Cavallo
Center for Public Archaeology
Rutgers University

RSVP

**RAINTREE
STECK-VAUGHN**
P U B L I S H E R S
The Steck-Vaughn Company

Austin, Texas

A production of B&B Publishing, Inc.

Editor – Jean B. Black
Photo Editor – Margie Benson
Computer Specialist – Dave Conant
Interior Design – Scott Davis

Raintree Steck-Vaughn Publishing Staff

Project Editor – Helene Resky
Project Manager – Joyce Spicer

LIBRARY OF CONGRESS CATALOGING-IN-PUBLICATION DATA

Willis, Terri
    Serengeti plain / Terri Willis
        p. cm. -- (Wonders of the world)
    Includes bibliographical references and index.
    ISBN 0-8114-6368-0
    1. Grassland fauna -- Tanzania -- Serengeti Plain -- Juvenile literature. 2. Serengeti Plain (Tanzania) -- Juvenile literature. [1. Grassland animals -- Tanzania -- Serengeti Plain. 2. Serengeti Plain (Tanzania)] I. Series.
QL337.T3W55 1995   94-3022
591.9678'27 -- dc20   CIP
                                   AC

Cover photo	Title page photo	Table of Contents page photo
Grant's zebras drink at a Serengeti water hole.	Yellow baboons and impala on the Serengeti.	Male leopard resting in a tree on the Serengeti

PHOTO SOURCES

Cover Photo: © 1986 Frans Lanting/Minden Pictures, Inc.

Animals, Animals. © Bruce Davidson: top 55
John Cavallo: 1, 3, 7, 9, 11, 12, 13 bottom, 14 both, 18, 20 top, 21, 26, 28, 31 both, 32 both, 33 top, 34 both, 35, 38, 40 bottom, 42, 45 bottom, 47, 60 bottom, 61
© Anne B. Keiser: 4, 10 top, 33, 36, 37 top, 40 top, 45 top, 50, 51, 52
© Dr. Alan K. Mallams: 8, 37 bottom, 41, 49, 58
FAO photo by F. Mattioli: 10 bottom
© W. Perry Conway: 13 top, 15, 16, 17, 22, 23, 27, 29, 30 bottom, 43, 46, 47 bottom

Photo by Gary Kramer: 20 bottom, 29, 54
© Fred Siskind: 24, 25, 30 top, 39
© Barbara von Hoffman: 25 top
National Fish & Wildlife Forensics Laboratory: 38 bottom, 56, 57
Theodore Roosevelt Collection, Harvard College Library: 48
UN Photo 146122 by O. Monsen: 59
UN Photo 153703 by John Isaac: 60 top

Printed and bound in the United States of America.
1 2 3 4 5 6 7 8 9 10 VH 99 98 97 96 95 94

Table of Contents

Chapter One

The Plain and the Fly

The tsetse fly is one of those ferocious insects that cannot be killed with a simple swat. You must actually squash this monumental pest between your fingers or under a shoe.

Attracted by movement, swarms of these large flies follow people as they walk across the grassy plains of the Serengeti in East Africa. The blood of humans, cattle, goats, and wild animals is the flies' only nourishment, and they pursue their food with gusto. Each fly has small, cutting teeth that can quickly slice open flesh and pierce a tiny blood vessel, so that a small pool of blood forms. The fly shoots a small amount of its own saliva into the blood to keep it from clotting, and then begins to suck up the meal, turning bloated and red.

The bite is painful for humans, but, even worse, it can be deadly. Some tsetse flies transmit a dangerous parasite called trypanosomiasis, the cause of sleeping sickness in humans. This parasite is picked up when a tsetse fly bites an animal or human with the disease and is then passed on to another victim by the saliva squirt.

Not everyone who is bitten by a tsetse fly will get the disease, but those who become victims of sleeping sickness face an agonizing death. The first symptom is excruciating pain in the neck, and then the discomfort grows as glands swell and the temperature rises. Finally, the body and mind deteriorate until death claims the victim.

Cattle don't contract sleeping sickness. Instead, they get an equally deadly disease called *nagana*. Wild animals, however, are incredibly immune to the deadly disease.

The vast open plateau making up the Serengeti is not the only place tsetse flies are found, but they are especially plentiful there. The flies find everything they need—plenty of shrubs and brush for their homes, plenty of animals for their meals.

Though the tsetse flies have caused untold misery over the years, we owe them a great debt of gratitude for one amazing achievement! They have kept human beings out of many regions of the Serengeti! Thanks

". . . Suddenly one sees the Serengeti, the plains stretching away to the horizon like the sea, a green vastness in the rains, golden at other times of the year, fading to blue and gray. . . . I shall never tire of that view, whether in the rains or the dry season, in the heat of the day or in the evening when one is driving down straight towards the sunset. It is always the same; and always different."

— Mary Leakey, anthropologist, in *Disclosing the Past*, 1979

A tsetse fly

The Serengeti Plain *(left)* is an open area of northern Tanzania in East Africa. Millions of animals live on and migrate over the plain as they have for hundreds of thousands of years.

to the awful threat of the tsetse fly, these areas have remained free from development and continued to be a haven for wild animals.

For centuries, humans have wanted to use the Serengeti, a part of East Africa in northern Tanzania that teems with wild animals and has a vast amount of open grassland. It could be a wonderful place to hunt for meat and trophies, a bountiful place to grow crops, and a mammoth pasture where huge herds of cattle could graze. But every time humans tried to take over a bit of the Serengeti, tsetse flies made them too miserable to stay in some areas. Almost uninhabitable, that is, for anyone or anything but millions and millions of wild animals: lions, giraffes, elephants, rhinoceroses, ostriches, cheetahs, hyenas, and many, many others. They live among the tsetse flies, and seem only slightly annoyed by an occasional bite.

Though new methods of controlling tsetse fly populations and treating sleeping sickness have been developed in the last half century or so, large sections of the Serengeti's vast wilderness remain virtually unaffected by humans. Life there carries on much as it has for more than a million years. There is nowhere else like it on the planet.

The Superb Serengeti

Most of the vast Serengeti Plain is now the Serengeti National Park. Established in 1941, it covers 5,700 square miles (14,763 sq km) along Tanzania's north-central border. It includes some of the best animal grazing land in Africa and is the only place in the world where massive land-animal migrations take place each year. The herds that travel across the wilderness include millions of wildebeests, gazelles, and zebras, making it an important tourist attraction for the country of Tanzania.

The park extends about 100 miles (161 km) east and southeast from the shore of Lake Victoria, with a strip that extends northward to Kenya. Though it is located beyond the plain, Africa's highest peak, Mount Kilimanjaro, dominates the view to the south. The Serengeti is one of the world's foremost national parks, located in a country with a number of extraordinary parks and wildlife. National parks take up more than 3 percent of Tanzania's land, and even more has been set aside for wildlife refuges. But the Serengeti, with its great diversity of plant and animal life, surpasses all.

"Never have I seen anything like that game," said a visitor of the Serengeti Plain in 1913. "It covered every hill, standing in the openings, strolling in and out among groves, feeding on the bottomlands, singly or in little groups. It did not matter in what direction I looked, there it was; as abundant one place as another."

Of course, other factors besides the tsetse fly can claim credit for this abundance. About 2.5 million years ago there was a widespread regional warming

On the vast open plains of northern Tanzania an estimated million and a half wildebeests live out their lives.

event that affected all of East Africa, including the Serengeti. This temperature shift resulted in the expansion of grasslands and the shrinking of woodlands. These grassland areas provide the food and space necessary for the large herds of animals, and the rainy season each year provides enough moisture to regenerate the grasses. The Serengeti's woodlands and mountain regions offer additional habitats to animals, and a network of rivers and lakes offers the water needed by the creatures that make the Serengeti their home.

Mount Kilimanjaro is visible from most of the Serengeti Plain. Here the famed mountain is seen through old acacia trees.

An Ancient Way of Life

Many people have compared life on the Serengeti today to life during the Pleistocene Epoch—the Age of Mammals. This period began nearly 2 million years ago. Paleoanthropologists, scientists who study the origins and evolution of humans, believe that the earliest human ancestors had their ancient roots in the Serengeti at about the same time. Throughout the world, the number of changes that have occurred since then have been nearly infinite. But on the Serengeti Plain, life continues to move to the same age-old rhythm that has been drumming all that time.

Similar patterns in life once moved through the vast spaces of Europe and America. Animals now known only by their fossils thundered across Europe. Herds of wild animals roamed the land freely, migrating regularly in search of food. Less than two hundred years ago, gigantic buffalo herds roamed the prairies of Canada and the United States. As recently as one hundred years ago, many places in Africa were host to great land migrations of animals.

The rhythm of life and death on the Serengeti continues much the same as always. This African buffalo carcass was food to other animals of the plain.

Humans altered these time-worn patterns by building villages and farms on the grazing lands, by limiting the movement of animals with mile after mile of fencing, and of course by killing the animals themselves. However, the Serengeti Plain remains to show us how nature ruled before humans took control.

The Nation of Tanzania

Tanzania, where the Serengeti is located, was formed in 1964 when the nation of Tanganyika joined with the island nation of Zanzibar. Both had recently declared their own independence from Great Britain. However, Tanganyika was a German possession for many years until after World War I.

Tanzania covers about 364,590 square miles (944,284 sq km), an area roughly the same size as the states of Texas and New Mexico combined. The residents do not have an easy living—life expectancy is only about 50 years—and most Tanzanians are very poor, with an average yearly income of $240 per person. The country's main export earnings come from agricultural products, particularly cashew nuts, cloves, coconut, coffee, and cotton.

Tanzania has a rather peaceful recent history, thanks mainly to the guidance of Julius Nyerere, a former schoolteacher who served as president of

People who come to Tanzania to view the wildlife at the Serengeti National Park and other parks contribute to the Tanzanian economy. Tanzanians recognize the economic importance of keeping such animals as giraffes safe.

Tanganyika and Tanzania for more than 30 years, until his resignation in 1985. While Tanganyika was establishing itself as independent of British rule in the 1950s, Nyerere's authority and wisdom kept the peace.

Nyerere's governing policy called for being open to the best ideas from around the world and providing health care, education, and enough food for everyone. But he felt that it would be difficult to give such care to all his citizens while they were spread out across the vast Tanzanian wilderness. As a result, most residents, particularly farmers, were gathered into small communities. This policy kept them away from much of the good land. The land they were allowed to use lay just outside these small villages, and its value was quickly depleted by overuse. While Nyerere's intentions were good, his methods often led only to worse poverty and eroded land. When he resigned in 1985, Tanzania's government was stable, and its people were literate but they were still very poor.

Nyerere can take credit, however, for helping to form Tanzania's excellent policy on wilderness conservation. The government, now led by Ali Hassan Mwinyi, is determined to maintain the country's national parks.

Although Tanzania is an underdeveloped nation, it is working hard to make sure that the Serengeti

A Tanzanian farmer hangs corn ears out to dry.

Beautiful waterfowl, such as these Egyptian geese, live alongside the more noticeable mammal population.

and other parks are protected, and that the needs of both its people and wildlife are met. This struggling nation is committed to preserving the beauty of its natural resources. Tanzania spends a higher percentage of its income on maintaining its national parks than the United States does on its parks. As one park warden explained, the Tanzanian government considers its wildlife not only a national heritage, but a heritage for all people.

The Serengeti is an important part of that heritage, not only for the plants and animals that live there, but for the benefit of human residents and visitors as well. People have long felt great curiosity about the wild animals of Africa. That curiosity draws many people to zoos. But zoos can't give us the pleasure that we get from seeing the magnificent animals in their natural habitats. In fact, there is satisfaction just in knowing that these habitats still exist—that in at least one small part of the planet, natural life goes on, untouched, as it has throughout the ages.

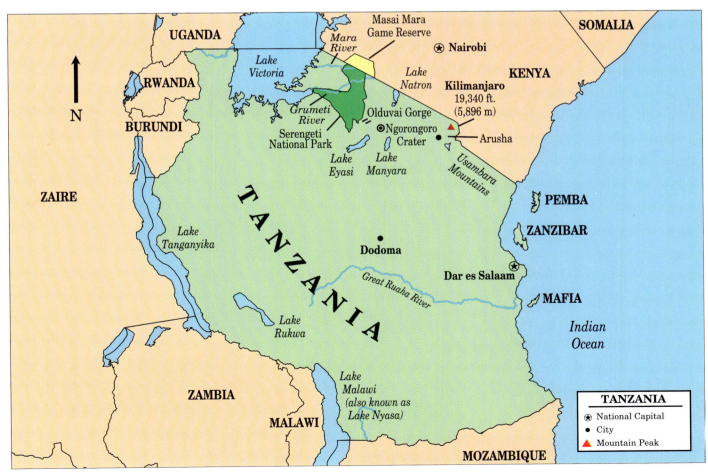

During the wet season the Grumeti River offers relief to water-loving hippopotamuses. However, in the dry season the river becomes a patchwork of small pools.

Chapter Two

A Vast Openness

The Serengeti is on a plateau that rises several thousand feet above sea level. The eastern part of the plateau is highest, about 6,000 feet (1,829 m), above sea level. It slopes toward the west to a low point of about 3,000 feet (915 m) near Lake Victoria. The rivers, too, drain westward into Lake Victoria, but even such major rivers as the Orangi, Duma, Mbalageti, and the Grumeti dry up during the long hot summers. Only the Bologonja and Mara rivers to the north flow throughout the year.

The oldest underlying rock, found in the western portion of the plateau near Lake Victoria, is from the early Precambrian Era. That period is the oldest (and longest) in the earth's geologic time scale. It was a time when the first primitive life forms—single cells—began.

The soil layer on top of the rock in the west is quite deep, so tall grasses and shrubs grow very well there. Dominated by red-oat grass, these plants rise to a height of about 3 feet (0.9 m). Trees don't grow here, though, because of the heavy alkaline (salt) content of the soil.

"It is a hard land, without the gentle charm and intimate views that give a person confidence and linger in his memory; instead, one feels that man is barely permitted to exist. It does not have scenic grandeur, but it has an austere beauty and the timelessness of great solitude."

— George B. Schaller, writing of the Serengeti in *Golden Shadows, Flying Hooves*, 1983

To the south, the soil layer is only about 3 feet (0.9 m) deep. At that depth, there is a thick, solid layer of calcium carbonate. This mineral is almost like rock—too solid for roots to penetrate. Only short plants with shallow root systems can grow here. This short-grass region of the plateau is known as the Serengeti Plain. Its name comes from *siringet*, meaning "the empty place" in the language of East Africa's Masai tribe.

The Surface of the Serengeti

The flatness of the Serengeti Plain is broken only by occasional outcroppings of granite stone called *kopjes*—an Afrikaans word meaning "little head." These rounded boulders are usually found in groups, providing at least some shade and shelter to the multitudes of animals grazing on the plains. The highest and largest group of kopjes, located near the center of the plain, is called Moru Kopjes.

Large acacias called "umbrella trees" provide much of the shade available to animals in the Serengeti.

Farther north, woodlands flourish. However, these forested regions are not like the dense forests of North America. Instead, they are grassy areas with widely spaced trees, especially types of acacia. One of the most common acacias is called the umbrella tree because it is flat-topped and wide, providing shade to animals that come to rest below. Acacia trees are actually a type of plant called a legume, related to beans and peas. Their nutritious beanlike seedpod is eaten by many animals.

The savanna, which links the regions of grassland and woodland, is a mix of woodland, grassland, and brush. It offers some of the most graceful scenery of the Serengeti. Here, too, kopjes add to the region's exotic beauty.

Rock outcroppings, called *kopjes* by the Afrikaans settlers, are features of the Serengeti. Sometimes the kopjes provide the only shade animals find. They are also used as dens by lions and leopards.

Animals Abound

The diversity of the savanna's plant life helps support the rich variety of animal life found on the Serengeti. Lions, elephants, giraffes, and gazelles mingle with huge herds of wildebeests and zebras.

About 25 species of hoofed animals—called ungulates—live in the Serengeti. Each has its own niche, or place in the ecosystem, which somehow works well with the others. For example, on the Serengeti Plain, several different species of ungulates eat different portions of the tall and short grasses. In doing so, some animals help other animals to more easily extract nutrients. This is called grazing succession.

The heavy grazers, such as the buffalos, hippos, and elephants trample and eat large, coarse grasses. Their digestive systems enable them to extract nutrients from the tough stems and branches that many other animals can't use.

Then, with the tough brush out of the way, the light grazers, such as the wildebeests, zebras, and topi, come in and make their meals from the grasses. This, in turn, makes way for the lightest grazers, such as warthogs and gazelles, to nibble on the tender, low-growing grass shoots and small herbs.

Animals have some harmful effects on the plant life, however, particularly during dry periods when edible vegetation is scarce. The plant life left by the drought is heavily overgrazed, which often damages the roots. But the animals encourage new growth

Waterbucks are a type of African antelope known for their unusually thick, shaggy coats.

Impala graze on grass and leaves and are known for their leaping ability.

by spreading seeds through their feces, loosening the hard soil with their hooves, and trampling the remaining plant life into a mulch that helps return nutrients to the soil.

Studies have shown that each animal species has a few favorite plants, and these plants are the most nutritious. Plants with the least food value go untouched as the animals move about to find their favorite plants.

The Seasons Spell Change

Weather is the basic reason for most travel patterns of the animals of the Serengeti. They simply go where there is water and grass.

The dry season typically begins in May, when all clouds disappear from the intensely blue sky. Images seen in the distance shimmer in the heat waves that rise from the ground. The wind blows steadily from the east, and brittle stalks of grass rattle softly.

Animals first gather near the few remaining water holes in the "corridor"—the narrow, low-lying strip of land near Lake Victoria. As even these wet

The primary grazing animals of the Serengeti are the wildebeests, also called gnus, zebras, and the long-horned Thomson's gazelles. They migrate to find water and the green grass that grows where it has recently rained.

NGORONGORO CRATER

Located in the highlands overlooking Serengeti National Park on the eastern edge of the plain is Ngorongoro Crater. Many travel writers suggest that visitors who have only a short time in East Africa spend that time at the crater, because all the riches of the region can be found in this one small place.

About 2.5 million years ago, the crater was a huge volcano, probably twice as high as the surrounding 10,000-foot (3,049-m) peaks that form the Crater Highlands Range. The volcano blew its top about a million years ago, and as its fiery interior cooled, it collapsed in the middle.

The bowl it formed is called a caldera. The Ngorongoro Crater caldera is the sixth largest in the world. Actually, it's the largest dry caldera. It is about 10 miles (16.1 km) in diameter, roughly 100 square miles (259 sq km) in all. Located at 5,600 feet (1,707 m) above sea level, it is surrounded by the rim of the crater, which rises another 2,000 feet (610 m).

The crater supports a great variety of wildlife, including the world's largest lion and spotted hyena populations. The highland forest on its tall rim helps catch clouds and hold moisture year-round, a rarity in East Africa.

Within these crater walls is a scenic area that many have compared to the Garden of Eden: a world filled with tropical vegetation, swamps and streams, lakes fringed with pink flamingos, birds of many colors flying overhead, and animals large and small roaming the land. The floor of the crater is similar to the Serengeti Plain, a large grassland with more than 25,000 grazing animals and patches and ribbons of woodlands. The most abundant animals are zebras, wildebeests, and gazelles.

Lake Magadi—the main lake—is home to beautiful flamingos, as well as spoonbills, stilts, avocets, terns, and the crowned crane, with its striking gold head. Hippos like to wallow in the lake, while rhinos and elephants wander into the water for a drink.

Ngorongoro National Park, which was established in 1956, is almost like a universe in miniature. Fortunately, it has been declared a World Heritage Site, which will help to preserve this amazing spot for future generations.

Shown in the picture above is a huge flock of flamingos in Lake Mayodi, located within Ngorongoro Crater. The crater rim is seen in the distance.

spots begin to dry, wildebeests, zebras, and gazelles begin their journey northward, in search of grazing areas in the woodlands.

Those plant-eating animals that don't migrate survive the dry period by eating whatever they can. Giraffes, for example, can nibble on the new leaves on the tops of trees, which retain their water. Other animals stay close to the rivers or the few other places that don't dry up.

The migrating animals will not return until November or December. Then rains will once again pour down on the dry plains, and the landscape turns from brown to green almost overnight.

As the land becomes drier and drier, swirls of dust, called dust devils, can arise in the wind. This Masai giraffe appears not to be bothered by the dust.

Chapter Three

The Great Migration

It is late in May, the weather is getting hotter, and the ground is beginning to dry to a crackled pattern of solid mud chips. What little water is left on the Serengeti Plain is rapidly evaporated by the sun and the grass will soon dry to the tawny color of a lion.

The animals feel their mouths begin to dry. Their nostrils burn as each breath pulls more dust into their lungs. Their throats parch. And yet, somehow, they sense that there is water somewhere. Perhaps they smell water, or, perhaps, a primitive instinct provides direction. Somehow, they know they have to head north.

And so the migration occurs. It is impossible to say exactly when the migration actually "begins," for it never really begins or ends. Rather, the movement of animals goes on and on in a circular pattern like the hands of a clock, stopping only for a few short weeks or months in areas where food and water are plentiful, then picking up again.

Wildebeests, zebras, and Thomson's gazelles travel 500 miles (805 km) every year during their continuous migratory journey. The map to the right shows the approximate migration route and indicates where the animals are located during different months of the year. During August and September, and from January to March, the herds pause to take advantage of plentiful food and water.

Day breaks on a May morning when the great wildebeest herds of the Serengeti feel the urge to move northward, toward water. The great migration begins.

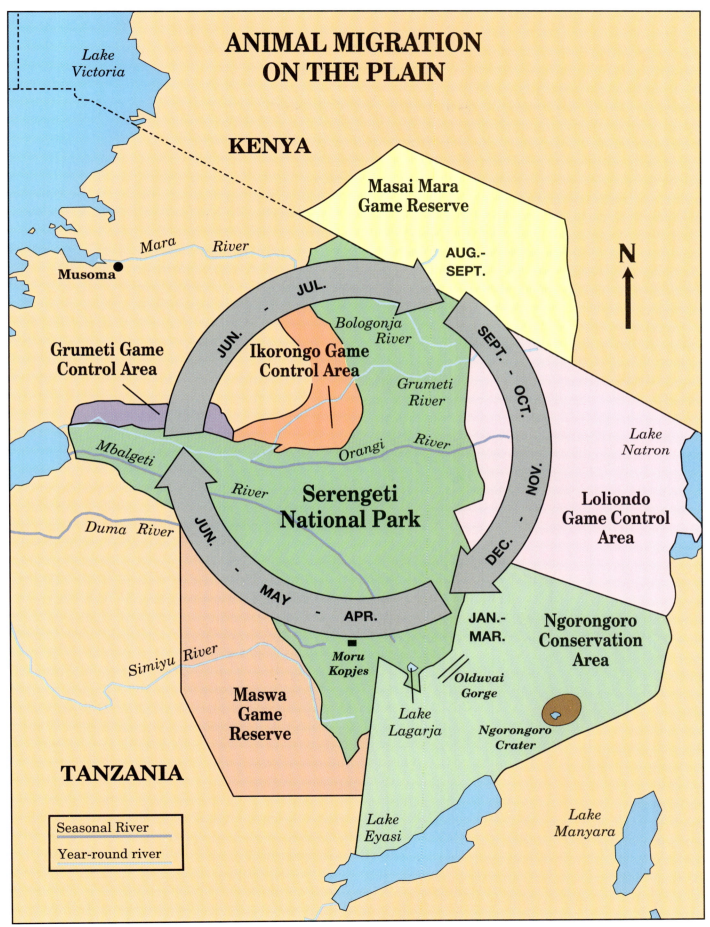

ANIMAL MIGRATION ON THE PLAIN

Lake Victoria

KENYA

Mara   River

Musoma

Masai Mara Game Reserve

AUG.-
SEPT.

N

JUL.

JUN.

Bologonja River

Grumeti Game Control Area

Ikorongo Game Control Area

Grumeti River

SEPT. - OCT.

Lake Natron

Mbalgeti

Orangi   River

Serengeti National Park

River

NOV. - DEC.

Loliondo Game Control Area

Duma River

JUN. - MAY - APR.

JAN.-
MAR.

Ngorongoro Conservation Area

Simiyu River

Moru Kopjes

Olduvai Gorge

Maswa Game Reserve

Lake Lagarja

Ngorongoro Crater

TANZANIA

Lake Eyasi

Lake Manyara

| Seasonal River |
| Year-round river |

Each year, the approximate route remains the same. The exact times of the migration vary slightly, because they are dictated by climate conditions, which can vary. More than a million and a half wildebeests, half a million Thomson's gazelles, and a quarter of a million zebras make the trek. Tagging along the edges of the herds and keeping very close tabs on the migrating animals are a number of other animals—the predators—including lions, wild dogs, leopards, and hyenas.

The Wild Wildebeest and Friends

Of all the exotic creatures that live in Africa, the wildebeest, or gnu, may be the most comical-looking. Alan Root, the famous wildlife filmmaker once remarked that they looked like they were designed by a committee. Even its name means "crazy animal" in the Afrikaans language. A wildebeest is an African antelope with a long, sad face and scraggly beard. It communicates by snorts and grunts. In herds, the wildebeests often buck and jump around for no apparent reason. Their feet seem to fly out in all directions as they run. But they can move swiftly when they have to.

As a species, wildebeests have adapted well to the demands of life on the Serengeti, as their large numbers show. They live in huge herds and feed almost entirely on grass. Because they need to drink often, they usually stay close to water sources.

More than a million wildebeests start the great migration that will take them across hundreds of miles of the Serengeti Plain.

Thomson's gazelles spend a lot of time eating, too—mostly grass. They drink water only when their grazing lands are dry and they can't get enough moisture from the vegetation. These small animals don't travel as far north as the wildebeests and zebras.

Graceful and quick, some "Tommies" can even outrun greyhounds. They have large dark eyes, long horns with ring-like ridges on them, and a dark, wide band along their flanks.

About half a million small, dainty Thomson's gazelles will accompany the wildebeests.

Their herds range in size from less than 10 to more than 500 animals.

Zebras, too, are social animals, active and noisy, always alert. They live in family groups of five to six females along with their young and a male, or stallion. Several family groups may herd together, but zebras always keep track of their own family members, perhaps through their markings, which are as individual as human fingerprints.

A major activity of the zebras' day, starting early in the morning, is eating. Grass, leaves, and bark are their main foods. Like the wildebeests, they need to drink regularly, and it is this need that propels them on their annual journey. They are always aware of possible danger from a predator while they eat or drink.

This zebra of the Serengeti is called Burchell's zebra. It is one of three species in Africa. Each animal has a slightly different pattern to its stripes.

A Major Migration

On their migration, the wildebeests, Thomson's gazelles, and zebras travel from 500 to 700 miles (805 to 1,127 km) each year. They move carefully too, not just helter-skelter in a general direction. They follow each other on well-marked trails laid down by generation after generation of migrating herds. These paths are so distinctive that they might be mistaken for paths made by humans. Only during feeding times, several hours each day, does the herd spread out to graze.

As the animals are preparing to leave the open Serengeti Plain, the wildebeests make the first

The need for water sends the great migrating herds on a year-long journey across the Serengeti of Tanzania and northward into Kenya. The wilde-beests, shown here, lead the way, along often invisible paths that they follow each year.

Although migrating zebras remain alert to possible trouble, the dust stirred up by the moving animals may prevent them from seeing a predator that might be following them.

move. Mating season has just begun, and the herd, numbering in the hundreds of thousands, is full of males leaping, bucking, and grunting to get the attention of females. Slowly at first, pairs form and head north. Then more pairs join them, and soon the whole herd is underway, moving six or seven abreast, raising a cloud of dust several miles long.

With them in that cloud are long lines of zebras. Though they join the wildebeests in the migration, the zebras keep to themselves in distinct groups of family units. The gazelles follow. First, the massive herd moves northwest, through the woodlands and savannas of the plain's western corridor. The animals stay here for a month or so and then continue north in June and July.

During this part of the journey, the herds pass outside the park's boundaries for a time. However, they remain protected from human hunters—the land they pass through is a special nonhunting region called the Grumeti Game Control Area.

Passing back into the park, they soon reach Tanzania's northern border with Kenya. Here the Serengeti National Park ends, but this, of course, means nothing to the animals. They continue north for a short distance, into a protected region of Kenya called the Masai Mara Game Reserve.

Unfortunately, the animals must first cross the Grumeti and Mara rivers, which are often torrents of deadly currents at this time of year. The migrating animals line up at the riverbank and throw themselves in as their turn comes. With all their

might, they swim against the raging waters, but many do not make it across. Thousands are drowned in the stampede across the river, and watchful vultures come in great flocks to pick at their carcasses. Others fall victim to large, hungry crocodiles. But many, many more are successful.

These animals spend most of August and September in the Masai Mara, a park combining huge stretches of rolling grasslands with forested hills and brush. They graze there, gaining fat and energy, preparing for their long journey south during September and October. Their instincts, built in each species over thousands of generations, tell them that the rainy season is near, and the plain will soon be able to provide food once again.

On the return portion of the migration journey, the animals again pass beyond the park boundaries. This time they are in a protected region called the Loliondo Game Control Area. Reentering the park near its far eastern boundary, they are back on the plains in November, in time for the rainy season.

Fresh new grass, which is particularly nutritious, springs up on the Serengeti Plain within a few days of the first rains. This season of bounty typically lasts until May or June. Just before the season ends, the wildebeest calves are born, mainly during a three-week period in February.

During the perpetual circle around the Serengeti, wildebeests give birth to calves that are ready to accompany the adults very quickly.

The migration is a matter of both life and death. Each step of the way, predators such as hyenas lurk to kill what they can, leaving the scraps for vultures.

Nature provides for the young animals. There is plenty of food available and fewer stalking predators than usual are around during this period. But the wildebeest calves by no means have it easy—they are expected to be up and walking within five or ten minutes of birth. Also, if a young calf is somehow parted from its mother, it is left behind. Many abandoned calves are seen wandering among the herds, approaching animal after animal, searching in vain for their mothers. Some will even go up to animals other than wildebeests. But usually the search is useless, and these young ones die.

Vultures are the clean-up animals of the Serengeti. They converge on dead animals and quickly pick the bones clean.

Protecting the Animals

When Serengeti National Park was created, it was not possible to include the entire migration route within the park boundaries. Political issues and other considerations also came into play. For example, since the animals cross into Kenya, the Tanzanian government had no control there. And some of the other land was important to native tribes, who had lived there peacefully for centuries. The government took these issues into consideration when drawing the final park boundaries.

In 1941, when the park was first established, its boundaries were quite different. It stretched out nearly twice as wide to the west across the southern portion and included all of the Ngorongoro Crater region. However, almost the entire northern half of the present park area—nearly everything north of Seronera (the only real village on the Serengeti Plain)—was excluded.

The boundaries were changed slightly in 1951, and then radically in 1959, when naturalists had a greater understanding of the animal migrations and the lands they travel through. The Masai wanted grazing land for cattle, also.

The planners solved both problems by creating the Serengeti Ecological Unit, an area larger than the national park. This 10,000-square-mile (25,900-sq-km) region, also known as the Serengeti Ecosystem, has natural boundaries that prevent the wildebeests from traveling any farther. The Loita plains of Kenya, which

When the Serengeti National Park was established in 1941, it included the territory colored pale green and dark green on the map below. However, in 1959, the park boundaries changed radically. The medium green and dark green territories highlight the present Serengeti National Park.

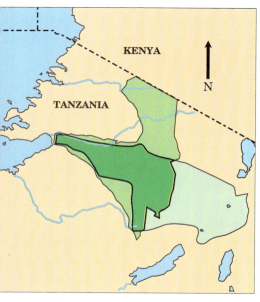

KENYA

N

TANZANIA

are very hot and dry and would not support the animals, serve as the northern boundary. The Loita Hills and the Gol Mountains are found in the eastern region. A very dry and rocky area, the Eyasi Escarpment, along Lake Eyasi, forms much of the southern boundary. In the west, the Maswa Game Reserve, where most land is cultivated, keeps the animals in.

Even though regions around the park are protected, this does not prevent people from killing some of the animals illegally. However, it reduces the large-scale operations that could kill thousands of animals at once during the migration.

A giraffe killed by a poacher and then left to rot. It will serve as food for such scavengers as vultures.

The Work of the Grzimeks

Many of the conservation regulations in force throughout the Serengeti ecosystem are due to the efforts of a father and son team—Bernhard and Michael Grzimek—in the late 1950s. Zoologists from Germany, the Grzimeks were called in to help the Tanganyikan government track the route of the migration and plan how to preserve the area.

The Grzimeks decided to divide the entire region into grids and then count animals within each section of the grid from a small, low-flying airplane. No one had ever tried to use an airplane for counting before. They practiced counting large herds until they became fairly good at it, and then each man made his own count, and they averaged the figures. Their unusual method of counting turned out to be quite accurate.

When the Grzimeks had to catch an animal to place a radio collar on it, they would drive alongside the galloping animal. Then, like a rodeo rider, either Michael or an assistant would jump on top of the animal, pulling it down. They would then rope the animal to keep it in place while they fitted it with a special radio collar to help them track it. Afterward, they let the animal go.

The Grzimeks found that there were many more animals than expected, and they followed a quite different migration route. These findings helped set the new park boundaries and the surrounding conservation areas.

Unfortunately, before he could return to Germany, Michael Grzimek died in a plane crash when a vulture struck and bent one of his plane's wingflaps while he was flying low. With their work in the Serengeti nearly complete, Bernhard finished it on his own. Michael was buried in a simple grave at the rim of the Ngorongoro Crater.

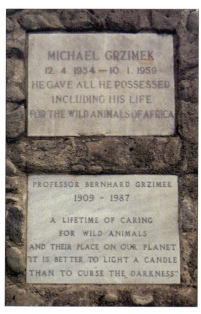

Though their work dealt with hard, cold facts, the Grzimeks faced the Serengeti with the wonder and curiosity of poets. Quite simply, they were in love with it. Unfortunately, his love for the land killed young Michael Grzimek, who is buried on the Ngorongoro Crater rim next to his father.

After his son's death, Bernhard wrote a book and produced a memorable film, both with the insistent title *Serengeti Shall Not Die*. In the book, he reminded us: "Men are easily inspired by human ideas, but they forget them just as quickly. Only Nature is eternal, unless we senselessly destroy it. In fifty years' time nobody will be interested in the results of the conferences which fill today's headlines. But when fifty years from now a lion walks into the red dawn and roars resoundingly, it will mean something to people and . . . they will stand in quiet awe as, for the first time in their lives, they watch twenty thousand zebras wander across the endless plains."

The Grzimeks' work had tremendous benefits for wildebeests. Since 1960, the number of wildebeests on the Serengeti has increased to 1.5 million. Now, a census is completed every other year, using the Grzimeks' methods, to count the wildebeests and to see if their migration pattern has changed.

Bernhard and Michael Grzimek succeeded in demonstrating that the great migration is important to the survival of the entire Serengeti ecosystem. Each animal that participates in the migration helps to sustain the bountiful diversity of plant and animal life that makes the Serengeti Plain such a special place.

Chapter Four

The Animals of the Serengeti

Though the migrating herds draw a lot of attention, they are only part of the more than 2.5 million animals living in the Serengeti. Each kind of animal plays its own part in the intricate pattern of life that has been woven there over centuries.

Every species of animals has its own place in the ecosystem—the food that it eats, the territory it controls, and the way it carries on its life. All these things work together to help maintain the natural balance that has supported wildlife for centuries on the Serengeti.

About 200 species of birds and more than 35 species of large mammals live in the national park. Elephants, giraffes, and hippopotamuses are found in large groups. Often, when visitors see so many animals at one time, they find it hard to believe that many of these animals are actually in danger of becoming extinct.

The Serengeti is home to a large variety of birds, including the magnificent crowned crane.

Among all the animals of the Serengeti Plain, some stand out as particular favorites among the tourists who come to see wildlife in action. Some are predators, and some are their prey.

Predators and Prey

Within the Serengeti ecosystem, there are both predators and prey. The prey animals are generally herbivores, or plant-eaters. The predators that eat them are, of course, carnivores, or meat-eaters.

Giraffes, even at their size, are prey. So, too, are the migrating wildebeests, zebras, and gazelles. Other prey animals include the tiny dik-diks, which stand no more than 15 inches (38.1 cm) high. The ugly tusked warthogs look ferocious but live on roots and grass. Ostriches, the world's largest birds, are also prey. Old and sickly birds are the usual targets, as are ostriches that are roosting—sitting on their nests of eggs.

Lions, hyenas, and wild dogs are predators, as are leopards and cheetahs. Leopards and cheetahs are major tourist attractions on the Serengeti. But they play an important role in nature, too, as they help to preserve the balance of the ecosystem. Cheetahs, the fastest land animals on Earth, can reach speeds of up to 70 miles per hour (113 kph), making them excellent hunters. They migrate along with the Thomson's gazelles, a favorite prey, hunting only during the day. Leopards, too, hunt the small gazelles, but they don't follow the herds and have a more varied diet that includes storks, pythons, and other antelopes. During the day, leopards are often seen resting in trees, which serve as good lookout posts for prey. Leopards may drag their kills, even large antelopes, high into the trees where they can feed without losing their catch to other predators. They do much of their hunting at night.

But the lines between predators and prey are not always so clearly drawn. For example, lions are predators, but decrepit old lions often become the prey of hyenas and wild dogs. Adult rhinos are rarely preyed upon—they are just too big for other animals to tackle—but occasionally they get stuck in the mud, and then the predators move in.

Vultures seem like predators as they hover over the migrating herds. But they are patrolling the

The strange protrusions on a warthog's face make it look like a creature to be feared, but it is basically quite timid and apt to hide from predators in burrows in the ground.

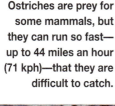

Ostriches are prey for some mammals, but they can run so fast— up to 44 miles an hour (71 kph)—that they are difficult to catch.

Big cats are the major predators of the Serengeti. The cheetah shown in the top photo has dined on a gazelle. In the bottom picture, a pride of lions can make a meal of one wildebeest.

WILDEBEESTS AND PREDATORS

The most common animals of prey are the wildebeests. As they travel the route of their annual migration, they encounter many animals that are eager to turn them into a meal. When wildebeests notice the approach of hungry predators, they quickly run away for an instant, stop, and look back to see if the threat has left. If they need to run at full speed, they tilt back their heads and throw their hooves up in a wild and frantic manner. Wildebeests provide up to 20 million pounds (9.1 million kg) of meat each year to their predators.

All wildebeest calves are born within several days of each other every year. This is nature's way of ensuring that enough wildebeest calves survive the ravenous appetites of lions and other predators. It takes only a short time for the young wildebeests to run fast enough to escape their predators.

Predators such as cheetahs, leopards, vultures, and hyenas, usually occupy a defined territory, but some of these animals move with the migrating wildebeests. It is difficult for wildebeests to get away from one predator, the cheetah. This animal is the fastest four-legged sprinter on Earth, sometimes running up to 70 miles per hour (113 kph). Usually a solitary hunter, a cheetah stalks prey in the early morning or late afternoon. Leopards are better climbers than cheetahs and often store wildebeest carcasses in tree branches.

Young male lions, unattached to a pride, sometimes follow the wildebeest herds all year, while lions in prides hunt animals only within their permanent home range. Wild dogs also stay within their extensive home ranges which are occupied by wildebeests much of the time.

The best thing humans can do to help the animals of the Serengeti survive is to understand the relationships that exist between predator and prey and not try to control them. For eons, the ecosystem has managed on its own, without human interference.

skies over the Serengeti looking for dead animals. As soon as they spot a carcass, they spiral downward toward their meal. They are the scavengers of the ecosystem. But lions, hyenas, and leopards also scavenge and frequently monitor the sky for signs of vultures descending toward a carcass.

Animals kill only for necessity. It is part of the life-and-death cycle that has ruled nature throughout time. In fact, predators help keep the ecosystem strong and healthy by keeping the number of prey down to the population that the environment will support. More animals die from starvation and disease when predators are removed from an ecosystem. Lions, for example, particularly enjoy zebras. During the average 10-year hunting lifespan of a lion, it kills about 190 head of prey, especially zebras with an average weight of 250 pounds (113 kg) each. This reduces the demand on the vegetation that zebras eat, and the remaining animals are assured of having enough food.

Predators also eliminate the weak and sickly, keeping the herds of prey animals hardy and alert. In this way, only the strongest animals go on to mate and reproduce healthy offspring.

Prey animals have, over generations, developed methods that have kept their populations strong. Some have developed good eyesight, fighting techniques, speed, or unusual running patterns that give them some advantage for avoiding predation.

Usually the males of the species are preyed on before the females. This often occurs when the males are alone and thus more vulnerable, such as when

The large predators, such as the leopard, are natural enemies of the gentler, plant-eaters—or prey—such as the gazelle being carried here.

Waterbucks eat aquatic plants while wading. They can leap through water gracefully and quickly when pursued by predators.

they are standing guard at the edge of a herd to watch over their females. In this way, they put themselves at risk for the benefit of the species. Since it takes only a few males to impregnate many females, the females are more important to the herd.

Wildebeests help protect their species by giving birth at a time when food is plentiful, and young calves are walking in about ten minutes, decreasing their vulnerability. Many are born within a short period of time, ensuring that predators cannot possibly kill enough of them to harm the species. Also, they are born during the months that are spent on the plains, at a time when many lions and leopards are in the woodlands. Other young, such as gazelles, crouch motionless and unguarded for several days after birth. They are almost invisible to predators while they gather the strength to follow the herd.

However, some activities put the wildebeests at risk, such as caring for young, going out to watering holes, and migrating. Death is a fact of life, and usually, if one member of the group is captured, the other animals will just calmly take notice of the commotion, and then carry on.

There are far more animals of prey than there are predators. Across the Serengeti Plain, there is approximately one predatory animal for each 100 prey animals.

African buffalo are strong animals that can weigh over 1,300 pounds (590 kg). If provoked, they can become ferocious and may even kill lions.

Lions are among the main predators that follow the migrating wildebeests. The male lion's roar sounds serious but probably means nothing more than an announcement of his presence.

Lions

The lion is nicknamed "King of the Beasts," and it's easy to understand why. The males have glorious and regal manes, and their roars send chills down one's spine. They can take down a wildebeest or zebra for a meal with just one massive bite. So perhaps they deserve their nickname.

On the other hand, perhaps they don't. Lions are

rather lazy creatures. They sleep about ten or fifteen hours each day on the average and doze for another one to five hours. They spend about one to five hours awake but just lying around, usually in a patch of shade under a tree or in the shadow of a kopje. Add all this up, and lions spend only one to seven hours each day being active. Many lions don't even eat every day but prefer to consume larger meals—as much as 50 to 70 pounds (23 to 32 kg) of meat—every few days.

Their roar isn't really so ferocious either. It is simply a method of communication. Many zoologists think it is also a way of showing off, like peacocks displaying their fine feathers. In the movies, particularly old Tarzan pictures, other animals always scurry off quickly at the sound of a lion's roar, but this doesn't happen in reality. Hunting lions like to sneak up on their prey, and they're not apt to announce their presence by roaring.

Despite their apparent ferocity, lions are affectionate within their family groups, called prides, especially with the young. Often, seven or eight cubs can be seen playfully romping among a group of elders, nibbling on the grown-ups' tails and jumping at their feet. The cubs get no punishment—only gentle licks on the face or back and a playful swat or two from a large paw. Male lions defend the pride's territory against intruders, keeping the young cubs safe. Prides consist primarily of related females, and lions

Female lions frequently hunt in groups. They sneak up slowly on their prey, usually selecting a wildebeest, zebra, or antelope that has wandered away from the herd. Then the hunters spring as one after the prey. Death comes quickly to the prey, usually by suffocation from a fast bite to the throat. The male lion then comes and feeds.

A lioness is usually about 8 feet (2.4 m) in length and weighs about 350 pounds (159 kg). The female lion is smaller and lighter than the male lion.

are the only large cats who live in female-dominated groups of this kind.

Females provide most of the menu for the pride's meals. There are no set rules about who eats first—the strongest members of the pride push the others out of the way and eat their fill first. Sometimes little cubs may starve to death, particularly during the dry season when prey is scarce. This practice may seem cruel, but it guarantees that only the strongest members of the pride produce more young.

"If you had to be reincarnated as a lion, the best place for you to be born would be in the Serengeti," wrote Bernhard Grzimek. "There is plenty of game, and large swarms of malarial mosquitoes and tsetse flies—as well as lack of water—make it uninhabitable by man. Even people who, at home, boast of their bravery as lion hunters admit that Serengeti is the place for lions."

Elephants

On the other hand, the Serengeti was not always the place for elephants. They had not lived within the national park for thirty years or more, until the 1950s. Then it became a refuge from the hunters who killed them for their precious ivory tusks. Today,

Elephants have a complex social organization and live in herds of all sizes, but their numbers are dwindling. They are in danger in Tanzania as elsewhere in Africa because of the demand for their tusks.

elephants live mainly in the northern savannas and woodland regions of the Serengeti. Here they can wreak havoc on trees, stripping them of their bark and uprooting them. At the same time, by means of their dung, elephants also disperse seeds, resulting in new tree growth.

Females are the backbone of the elephant social order, and the family group, or herd, is made up of mothers, daughters, and younger sons. Older males, or bulls, are expelled from the group at puberty, which, like humans, occurs at about age 13. The solitary bull elephants join loosely organized herds that constantly break up and rearrange themselves. A male will join a group of cows only if one female is ready to mate and receptive to him, and then only for a few days.

African elephants are the world's heaviest land animals and the second tallest, next to the giraffe. When they flare their enormous ears in alarm, they seem even bigger! But like the lion's roar, the flared ears and raised trunk of an elephant aren't as ferocious as they seem. When they have this posture, elephants aren't angry, they are curious and excited. They want to use their full senses of hearing and smell in exploring something.

African elephants have sought refuge within the protected areas of the Serengeti. They have long attracted poachers who kill the huge animals for their tusks.

An elephant scratches its trunk on a dead tree.

An elephant may use its trunk to trumpet a warning to other members of the herd. But an elephant that is truly angered will actually pull its ears in and its trunk down while charging. A charging elephant can reach speeds of up to 20 miles an hour (32.2 kph) for a short period, but typically they walk at about 3 to 4 mph (4.8 to 6.4 kph), about the same as people.

Their trunks are strong enough to carry heavy objects and pull down large trees. But they can also make the delicate motions required to pick up and carry small objects like coins.

In zoos, elephants can be seen sleeping lying down. This is because they are calm and have little to fear, so they can relax. In the wild, however, elephants sleep standing up so that they are always ready to move if need be.

Rhinoceroses

While elephants may be alert and cautious sleepers, rhinoceroses are quite the opposite. Since they are most active at night, they sleep during the day, either while lying in the sun or wallowing in the mud. They sleep soundly but can be on the move—charging—if they are awakened abruptly.

Rhinos have a sort of security system to warn them of possible trouble while they're sleeping. Tick birds, also called ox-peckers, like to rest upon the backs of the dozing animals. When an intruder approaches, the birds fly around making a noisy racket, warning the rhinos. It also warns bystanders who do not want to anger sleepy rhinoceros.

Actually, it is difficult for a charging rhino to cause that much trouble because these animals have very poor eyesight. Often they will run toward the intended victim and then suddenly stop several feet away because their eyes are unable to focus directly on the target. Amazingly, a charging rhino can run as fast as a galloping horse, but not for long.

Basically, though, rhinos are peaceful animals. They love mud baths and will grunt happily in the

Because rhinoceroses are active at night, they use the daytime hours to rest in the warm sun.

A carved rhino horn that was confiscated by the U.S. Fish and Wildlife Service because it is illegal to sell parts of endangered animals in the United States.

muck for many hours. Young rhinos, which weigh between 45 and 85 pounds (20 to 39 kg) at birth, remain with their mothers for about three or four years. Then they usually live in the same limited area for life, dining mostly on twigs, small shrubs, and young trees. Because a grown animal can eat as many as a hundred small trees each day, rhinos have a profound effect on the ecosystem. If these endangered animals were to become extinct, the landscape would probably be taken over by heavy tree growth.

Rhinos are in danger of becoming extinct before the year 2000, not only on the Serengeti but throughout Africa and Asia. Poachers often kill the animals just to cut off their valuable horns. Made up of compacted, hairlike tissue, with a texture and hardness similar to human fingernails, rhino horns are prized as medicines in Asian countries, particularly Taiwan and China, where they are ground up and used for lowering fever and blood pressure. Many people also believe that ground rhino horn is an aphrodisiac, or "love potion." In the small Middle Eastern nation of Yemen, nearly every adult male carries a dagger, called a *jambia*. The possession of a dagger with a rhino-horn handle confers great status on the owner. This small country is one of the main buyers of horns from illegally killed rhinos.

Penalties against poaching are very severe, but

The African black rhinoceros, sometimes called the two-horned rhino, lives a solitary life, except when a mother has a youngster, which stays with her. Oxpeckers, also called tick birds, ride on a rhino's back while removing maggots and ticks from its hide. If danger approaches, they make a hissing sound to warn the rhino.

poachers are rarely caught. Unfortunately, sleeping rhinos make very easy targets for hunters. In the Serengeti Plain as in all of Africa, the number of rhinos has decreased by about 90 percent during the last century.

Giraffes

Giraffes are the world's tallest land mammals, reaching heights of up to 20 feet (6.1 m) and weighing as much as 2,600 pounds (1,179 kg). Their height lets them browse for leaves high in the trees, where no other animal competes with them for food. Their long legs enable them to run fast and gracefully, but they cannot run very far because they are short-winded. Their two hearts work together to pump blood 7 feet (2.1 m) to their brains!

They range in color from pale orange to dark gray. As with zebras, each giraffe's markings are unique—no two giraffes have the same patterns.

The tall, long-legged giraffe looks as if it would be clumsy when it runs, but it moves with speed and grace. It fends off possible predators with powerful blows of its feet.

No other mammal competes with the giraffe for the tender leaves at the tops of trees. Only the giraffe can reach them.

They travel in small groups, which makes each individual less vulnerable to lions. If they are attacked, they can be fierce fighters, kicking with powerful hind legs, but, again, they tire easily.

It happens rarely, but if two male giraffes decide to fight each other over a female, they don't kick or bite. Instead, they stand side by side, facing the same direction, and with wide swings of their necks, they smash their skulls into the other's head, neck, or chest. Rarely do these fights end in injury to one of the animals. Once the winner is determined, they resume living peacefully together.

Hyenas

Hyenas, also called spotted hyenas, live in social groups called "clans." With up to a hundred members in a pack, they share their food. Mostly nocturnal animals, they spend their days in a dark maze of holes and underground passageways. Females are the larger and dominant animals of the species.

Hyena territories usually cover about 100 square miles (259 sq km). They rarely hunt outside their own territories. On the Serengeti, however, things are sometimes different. Here the packs will often follow the migrating herds of hoofed animals—their prey—across the plains.

For a long time, people thought hyenas were cowardly scavengers, eating only carrion, or meat left over from the kills of other animals. This is not true. Hyenas hunt in packs, sometimes chasing their prey at speeds of up to 40 miles per hour (64 kph). Their powerful jaws and sharp teeth can smash through bones easily.

Researchers have learned that lions often come after prey hyenas have just killed, forcing the hyenas to wait nearby. This image led people to think the lions had made the kill and the "cowardly" hyenas were waiting for the remains.

Wild Dogs

Another group of successful hunters are the wild dogs, considered to be among the fiercest predators in all Africa. Wild dogs, the African equivalent of North American coyotes, are fearless, daylight hunters, going after animals much larger than they

There is nothing really cowardly about the hyenas that follow the migrating herds, hunting in packs.

are, such as wildebeests and antelope. Able to reach speeds of up to 35 miles per hour (56 kph), they can run down most creatures. They usually hunt in the early morning and evening hours, always in packs of about five to ten animals. One dog gets a grip on the prey, while the others tear at its flesh. However, they can act in friendly ways to get what they want, such as licking another dog's face to get a piece of the kill.

All members of the pack take care of the young by regurgitating food onto the ground for the babies after a hunt. They also use this method to feed new mothers who can't leave their pups, as well as sickly dogs who can't join the hunt.

When pups are old enough to follow along on a kill, they are the first to feed. Adults circle around, protecting them until they have eaten their fill. Only then do the adults eat. This assures the survival of

Africa's wild dogs hang around the migrating herds, waiting for the opportunity to kill and feed on injured or isolated animals.

the species, and the young enjoy this special treatment for about eight months. This species is the most endangered carnivore in the Serengeti, though, due in part to fatal outbreaks of rabies and distemper.

Anything that affects the population of one species on the Serengeti can affect the populations of other species. The relationship among predators and prey is an intricate one. Doom can come to the wildlife of a region if outside influences change the way in which the two groups of animals live and interact. Humans must tread carefully before interfering.

Chapter Five

Humans on the Serengeti

Animals, for the most part, live untouched by humans on the Serengeti, but humans have left their mark. Some people came to kill big game, so that they could show off their trophies and brag of their courage when they got home. Others came with opposite intentions, to help preserve the wildlife and keep it free from further harm. Many more came just to view the magnificent land and its creatures.

Still other inhabitants of the Serengeti were there not by choice but by the grand design of evolution. They were the ancestors of modern human beings, the earliest people to walk on the Earth.

Olduvai Gorge

These first humans evolved in an area called Olduvai Gorge, which lies in the southeastern portion of the Serengeti Ecological Unit. Today, the gorge attracts many visitors as a world-renowned archeological research site known as the "Grand Canyon of Evolution."

People come from around the globe to the site where our earliest ancestors lived, worked, and died. Here, in this dry and desolate gorge, archeologists and anthropologists Louis and Mary Leakey unearthed two teeth and parts of the skull of a primitive human they called *Zinjanthropus boisei*, or "East Africa man." It lived about 1.75 million years ago.

Many millions of years ago, Olduvai Gorge was the site of Olmoti Volcano whose lava covered the region. Shallow "Lake Olduvai" formed, its shores lined with papyrus swamps and trees. The lake and nearby waterways attracted animals such as jackals, hyenas, elephants, flamingos, and a huge kind of wildebeest. The size of this lake fluctuated as the climate changed. Other volcanic eruptions added layers of ash to the land.

The lake was drained about 1.5 million years ago when the Earth's crust shifted. A river still flowed through the open, grassy savannas, however. The region's climate changed repeatedly as the years went by, becoming much drier, and then wetter.

About 100,000 to 30,000 years ago, a series of severe earthquakes changed the landscape completely. After this disturbance, a river, which ran during the wet seasons, cut down through seven distinct

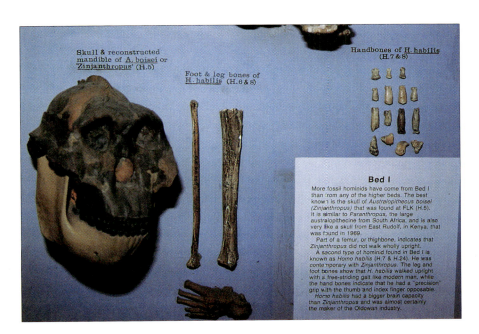

Skull & reconstructed mandible of A. boisei or 'Zinjanthropus' (H.5)

Foot & leg bones of H. habilis (H.6 & 8)

Handbones of H. habilis (H.7 & 8)

Bed I
More fossil hominids have come from Bed I than from any of the higher beds. The best known is the skull of *Australopithecus boisei* (*Zinjanthropus*) that was found at FLK (H.5). It is similar to *Paranthropus*, the large australopithecine from South Africa, and is also very like a skull from East Rudolf, in Kenya, that was found in 1969.
Part of a femur, or thighbone, indicates that *Zinjanthropus* did not walk wholly upright.
A second type of hominid found in Bed I is known as *Homo habilis* (H.7 & H.24). He was contemporary with *Zinjanthropus*. The leg and foot bones show that *H. habilis* walked upright with a free-striding gait like modern man, while the hand bones indicate that he had a "precision" grip with the thumb and index finger opposable.
Homo habilis had a bigger brain capacity than *Zinjanthropus* and was almost certainly the maker of the Oldowan industry.

Many of the Leakeys' finds can be seen in the museum at the visitor's center near Olduvai Gorge, where people come to learn about human beings' ancient past.

43

DIGGING UP THE HUMAN PAST

Anthropologist Louis S. B. Leakey was born in Kenya. His parents, British missionaries, raised him among the Kikuyu people with whom they worked. After attending a university in England, he began to explore Olduvai Gorge in 1931.

Two years later, Leakey married Mary Nicol, an English anthropologist who was very skilled at making drawings of archeological finds still in the ground. Together, the pair spent almost 40 years digging up the past in Africa, especially in Olduvai Gorge. They worked carefully over the years, gathering information about ancient life. From the tools and animal bones they found, they knew that early humans ate meat, using tools to capture and prepare frogs, lizards, rodents, and even pigs, giraffes, and elephants. Mary continued to dig after Louis's death in 1972. Among her discoveries was a set of fossilized human-type footprints made 3.5 million years ago.

Mary and Louis Leakey's son Richard, born in 1944, rejected the "learned" life and became a safari guide . . . until he found his first human fossils and decided to become an anthropologist, too. In 1967, he discovered a human-fossil bed, called Koobi Fora, in Kenya, that has turned out to be even richer in finds than Olduvai Gorge. He has founded the Louis Leakey Memorial Institute for African Prehistory.

Australopithecus boisei

Homo habilis

Homo sapiens

layers of sediment and rock creating the deep ravine and side valleys we now call Olduvai Gorge.

Finding fossils is fairly common in this gorge, and it is easy to determine how old each fossil is by the time period of the layer it came from. Judging from the bones they've found, scientists studying the area know that some amazing creatures once roamed the land there: giraffes with large antlers, elephants with tusks that grew downward from their bottom jaws, and hippopotamuses with eyes that projected from their heads like periscopes.

But the most important discoveries are those that concern early humans and human-like ancestors. Louis S. B. Leakey began his search in 1930 when he found some crude stone tools at Olduvai Gorge on his first visit there. At that time he was the lone voice who proclaimed an East African origin for our earliest ancestors.

It wasn't until July 17, 1959, that Mary Leakey found the *Zinjanthropus boisei* skull. The Leakeys felt that it was a totally different sort of prehistoric human, but further research showed that it was actually an *Australopithecus*, a rather common fossil of early humans. The name has since been changed to *Australopithecus boisei*, though many anthropologists still refer to this skull as "Zinj." Zinj was important because it supported the Leakeys' theory that humans had evolved from an African ancestry.

A year later, the Leakeys discovered another skull, also dated to about 1.75 million years old, which is directly in the human line. They put it in the same genus classification as humans—*Homo*—naming it *Homo habilis*, or "handy man." Handy man had a large brain, but was small in stature, probably only about 4 feet (1.2 m) tall and weighing 75 pounds (34 kg). *Homo habilis* lived in the Olduvai region for about 500,000 years, using the same kind of primitive tools for the entire time.

But even though human history may have had its beginnings near the Serengeti Plain, it is impossible to directly connect these early humans to the people who live there now. For thousands of years, Africa was a crossroads, with many wandering tribes traveling over the continent before finally settling. Usually these travels can be learned about only through the tribe's language and the many stories handed down through the generations.

The Masai

There are many native tribes living in Tanzania, but one conspicuous group in the Serengeti is the Masai. They are probably descendants of North African tribes who traveled south along the Nile to the East African region about 400 years ago.

In the 1800s, the Masai were known as fierce warriors to the early Colonial explorers. Their reputation helped keep the Serengeti wild, because many explorers were afraid to enter Masai territory. At that time, there were about 50,000 Masai living in the area.

Today, fewer than 20,000 Masai live in the Serengeti Ecosystem. And though they still hang on to their culture, they are no longer warlike. Until very recently, however, Masai warriors were expected to

Masai beadwork patterns are handed down from generation to generation.

A young Masai girl leans against the wall of her family's sturdy hut made of sticks and cow dung.

prove their bravery by killing a lion with a spear. Now they are nomads who keep cattle, sheep, goats, and donkeys, and move with the herds in search of good grazing.

The Masai believe that their God created all the cattle in the world for them to care for. They are proud of their large herds of cattle, and everyone takes part in caring for them, even the young children. Women handle most of the milking chores, while men are in charge of the cattle. They don't believe in traditional fences. Instead, they surround their cattle with uprooted thornbushes that keep the cattle in and the predators out.

The Masai live almost entirely on milk and blood from their cattle. The blood is obtained by lancing an animal's jugular vein and drawing it out. The Masai live in large, extended families, and each man takes several wives. Each wife has her own home and herd of cattle. Children are very important in the culture. The elderly are protected and revered. Though it is not always easy, the Masai are trying hard to maintain their age-old culture and life-style in today's modern world.

MASAI IN NGORONGORO

The beautiful Ngorongoro Crater and much of the land surrounding it, a total of 3,200 square miles (8,288 sq km), make up the Ngorongoro Conservation Area. It includes all of the crater highlands and part of the Serengeti Plain, too. It was established to help meet the needs of the native Masai people.

The herders were told in the mid-1970s that they could no longer graze their cattle and goats in the Serengeti National Park. Since this was land they had been using for generations, they naturally felt a good deal of resentment about the new law. To show their anger, some Masai turned to banditry—spearing the wild animals—something they would normally never do.

In response, the government granted the Masai the right to live and raise their cattle outside the Ngorongoro Crater, and to bring the animals inside the Crater to drink from streams within. However, they are not allowed to live inside the crater, where many had previously made their homes, nor are they allowed to grow crops within the conservation area.

So far, the plan has been workable, though not perfect. There is still a good deal of resentment on the part of many Masai, but most view the regulations of the conservation area as the best way to combine the needs of native people with the protection of the priceless wildlife of this natural wonder.

Scientists on the Serengeti

The Masai are not the only people who travel across the Serengeti. The vast abundance of wildlife gathered there—the greatest concentration of animals on the planet—makes the area a magnet for all types of researchers. The Serengeti Wildlife Research Institute at Seronera in the national park is the area's most impressive study center. It hosts wildlife biologists, experts on animal behavior, and many other scientists who come from Tanzania and all over the world and stay for a number of years.

A wide variety of topics are studied in great depth, from the social habits of the lion (the "king of the beasts"), to the life cycle of the lowly dung beetle. Although small, the dung beetle plays a very critical role in the survival of the Serengeti. Rob Foster, a biology researcher from Oxford University, says, "There are two million herbivores on the Serengeti plains; that's [thousands of tons] of dung every day. The dung provides nutrition to plants, and the main agent of its burial [putting it close to the plant roots] is the dung beetle. It is the critical link."

In addition, other research projects focus on such animals as spotted hyenas, wild dogs, cheetahs, and lions. All these studies make the Serengeti one of the best-studied ecosystems in the world.

Olive baboons live in groups called troops, usually in rocky, open country. Their main enemies are leopards.

An African wildlife biologist working in the Masai Mara of Kenya

Hunters on Safari

While the intent of the researchers is to save and protect the animals, not everyone who travels to the Serengeti agrees. Some people come to hunt.

For generations, the Tanzanian area has held a reputation as the ultimate safari destination. In fact, the word *safari* comes from the local Swahili language, meaning "a trip." Today, many people are still drawn by the call of an exotic Serengeti safari. But now they must do their shooting with cameras, not guns. Hunting is illegal throughout the park, but not in some adjacent areas.

Until about 1930, safari hunters spent several weeks on horseback and on foot in the wilderness,

tracking the prized animals. They had to take all their provisions for the safari with them, and they camped each night at a convenient spot.

That may sound simple, but such safaris were certainly not primitive. Most hunters hired large groups of Africans to carry all their goods. A few carried the guns and served as guides to the area and its animals. All the white hunter had to do was walk along and fire the weapon when they came upon game. Many hunters took as much pride in beating their servants to "encourage" them to do their work as in their marksmanship. Frequently, all the ingredients of civilization were brought along—fine china and silver, excellent foods, and wines. Even chefs and valets came along to make their employers feel at home.

More recently, hunters began to use motor vehicles to get around. They made a base camp in one spot, then drove around looking for game. When animals were spotted, the men and women jumped out of their Jeeps or Land Rovers long enough to shoot. They were back in their comfortable camps in plenty of time for cocktails before dinner.

In the first half of the twentieth century, hunting safaris were the rage for Europeans and Americans. Among those who enjoyed safari hunting was Teddy Roosevelt, who left Washington for an African safari just after finishing his term as president of the United States.

Another American, writer Ernest Hemingway, also loved safaris. "Now, being in Africa," he wrote, "I was hungry for more of it—the changes of the seasons, the names of the trees, of the small animals, and all the birds."

"Of all dangerous beasts the lion, when once aroused, will alone face odds to the end. The rhinoceros, the elephant, and even the buffalo can often be turned aside by a shot. A lion almost always charges home. Slower and slower he comes as the bullets strike; but he comes, until at last he may be just hitching himself along, his face to the enemy, his fierce spirit undaunted. When finally he rolls over, he bites the earth in great mouthfuls; and so passes, fighting to the last. The death of a lion is a fine sight."

—Stewart Edward White, big-game hunter, 1927

But while Hemingway appreciated the beauty of nature, he also enjoyed the kill. In his book *Green Hills of Africa*, he wrote of the "great humor" to be found in watching a hyena staggering, flopping, and tumbling to its death after being shot by a hunter's bullet.

Many other hunters, in fact, took great delight in writing books and articles about their feats. This no doubt helped add to the worldwide enthusiasm for safaris. Both men and women participated and killed thousands of animals over the years, including elephants, giraffes, and zebras.

Rhino hunting was also popular, especially in the 1920s, when their value increased. They were numerous then and easier to kill than elephants. Some hunters bragged of shooting more than a hundred rhinos a year for several years, mostly for their horns. But others shot the rhinos just for the "fun" of it.

Stewart Edward White, a big-game hunter in the early 1900s, wrote several books on safaris. In one, he described rhinos as a "nuisance," in part because they could hide so well in the dense scrub and remain unseen until they came charging on an unwary hunter. It was his opinion that all rhinos must be wiped out, in order to help civilize Africa and make it a place where humans can live safely.

Lions eyeing marabou storks, though the birds will probably not serve as a meal.

"No such dangerous lunatic can be allowed at large in a settled country, nor in a country where men are traveling constantly," White wrote of the rhinoceros. "The species will probably be preserved in appropriate restricted areas. It would be a great pity to have so perfect an example of the Prehistoric Pinhead wiped out completely. Elsewhere he will diminish, and finally disappear."

Though his prediction so far has not come true, he may eventually be correct. Rhino numbers are still dropping. Poachers who kill for precious rhino horns continue to deplete the species.

It was the majestic lion, however, that was the most prized trophy to hang on the wall back home. Every safari hunter, it seemed, had to be able to brag of his or her victorious battle with a lion. White wrote of stalking and shooting down a lion, and ended by saying, "This is real fun."

One of the most exciting ways to view the animals of the Serengeti is by hot-air balloon *(right)*. However, the noise of the propane burner used to heat the air can scare the animals as the balloon passes by overhead.

Safari travelers find that waterholes—even little puddles like the one shown—are good places to find animals to observe.

Modern Safaris

Before the safari hunts could choke out the lions and other animals on the Serengeti forever, the national park was created. As people around the world became more familiar with the amazing wildlife in this unique ecosystem, public opinion changed, and trophy hunting was frowned on. The last shot by a safari hunter within the park boundaries was fired in 1951. Since then, park laws have prohibited hunting, and the populations of most species have built up again.

Today, visitors come just to appreciate and photograph the sights. For many years, there was little tourism in Tanzania. For several years after independence, tourists who did enter were met by a poor economy. In recent years the economy improved enough to provide first-class tourist facilities.

Game drives, in which tourists travel together in large Land Rovers and mini-buses led by knowledgeable guides, are the standard mode of travel for modern safaris. Usually, the visitors will encounter lots of wildlife because most animals have lost their fear of cars. Also, the drives are far safer than foot travel, which isn't even allowed in most parks. Most game drives are made in the early morning and late afternoon hours, when the animals are most active. Unfortunately, nighttime drives are not allowed in Tanzanian national parks, so lions, leopards, and hyenas on the hunt are rarely seen.

Some tourists, accustomed to the noise and congestion of modern cities, are stunned by the wide-open space and solitude that the Serengeti offers. It may be an uncomfortable feeling, having to suddenly adapt to an environment where normal life—among both people and animals—so closely resembles life as it must have been a million years ago.

Some very lavish tourist lodges have opened on the rim of the Ngorongoro Crater, and more are being built in and around the Serengeti National Park. Tourists can find all the amenities at such places as Sopa Lodge in Ngorongoro Crater.

Chapter Six

Looking Toward the Future

People in Tanzania and throughout the world hold the key to the Serengeti's future. Will it survive as a beautiful wilderness and safe haven for animals—a marvel for generations to behold? Or will humans destroy one of the planet's last great natural environments? There are people working toward both outcomes.

People have had an influence on the Serengeti for hundreds of thousands of years. Prehistoric humans were predators and scavengers, much like the lions and hyenas are today. They used simple stone tools and relied on cunning to survive and to obtain the meat they needed. These early humans also fell prey to some faster, stronger predators, so they were all part of the food chain.

But as humans evolved, they learned to use more powerful weapons—such as spears and later arrows they could shoot game with.

Today, humans have guns and traps and snares. They can do a lot of damage to animal communities if they choose. Unfortunately, many people are doing just that. Poaching is by far the biggest problem confronting Serengeti National Park.

Poaching

Hunting has some benefit for the animals. In areas or during periods where predators are scarce, hunters keep populations under control, reducing starvation and disease. Some game reserves near the Serengeti are open to hunters but only at certain times, and specific licenses are required. Poaching—hunting illegally—is different.

While Tanganyika was still a British colony, many hunters came to the Serengeti from the older British colony of Kenya to slaughter the lions, which were considered dangerous pests. Some hunters killed 100 or more on a single safari trip, hacking off only their tails as trophies.

In 1929, hunting was prohibited in 90 square miles (233 sq km) of the open plains toward Lake Victoria. Since that time, and since the national park

"In the coming decades and centuries men will not travel to view marvels of engineering, but they will leave the dusty towns in order to behold the last places on earth where God's creatures are peacefully living. Countries which have preserved such places will be envied by other nations and visited by streams of tourists. There is a difference between wild animals living a natural life and famous buildings. Palaces can be rebuilt if they are destroyed in wartime, but once the wild animals of the Serengeti are exterminated no power on earth can bring them back."

— Bernhard Grzimek, in Serengeti Shall Not Die, 1961

Elephant poaching is still a problem on the Serengeti. These animals are killed for their valuable ivory.

The giraffe's main predators are lions and man. Although they are protected in East Africa, they are still threatened.

was established, things have improved somewhat. But poaching is still a major destructive force.

Rhinos have been poached almost to extinction for their horns. Elephant tusks are also prized items to poachers. The ivory from tusks brings a fine price in many countries. More than half the park's elephants have been killed in recent years, and only about 500 remain.

Poaching is now actually an industry of sorts. Huge, illegal organizations take care of all the details. There are people who set snares, others who butcher carcasses and deal with the remains. Some are in charge of hiding forbidden cargo while it is taken into town, and still others do the actual selling—to people who don't care that the items they are buying were illegally obtained.

Some poaching is done on a smaller scale. Individuals may kill a wildebeest or zebra now and then to feed their hungry families. Poor peasants may have a small operation—killing a few animals at a time. They butcher them out in the wilderness, cut the meat into thin strips, and hang it on a line to dry before it is sold at the local village market. About 40,000 to 50,000 animals are killed each year by meat poachers.

More killing is done for souvenirs. Many times poachers just cut off parts of the dead animals, such as lion paws, for trophies. Zebra tails are used as flyswatters. Sometimes only the hides are taken. Elephants' feet are made into wastepaper baskets. These items are primarily sold to white tourists.

Today most poachers are Africans who have been hunting game all their lives and don't understand the need to stop. And perhaps we should not be too quick to judge them. Most local people earn only a few hundred dollars a year. By poaching, they can more than double their annual income in just about three months. It's easy to understand how attractive poaching can be.

In any case, this illegal practice goes on despite the efforts of the Tanzanian government, which employs about 85 antipoaching rangers to police the park. Even in the protected Ngorongoro Crater rhino poaching is a problem. The crater covers only 100 square miles (259 sq km) and is protected by natural enclosures, but still the illegal activity continues. If it is so difficult to control in this confined area, it's easy to see that eliminating poaching on the wide-open Serengeti is nearly impossible. As long as there is a market for illegally killed animals, the crime will continue.

Some steps are now being taken to decrease the demand worldwide. Leading this global campaign is Richard Leakey, the son of Louis and Mary Leakey and the director of Kenya's Wildlife Conservation and Management Department. He wants to stop the widespread poaching of elephants and rhinos throughout East Africa.

In the United States, an inspector looks at a carved ivory decoration that is illegal to own if it was made from the tusks of a recently killed elephant.

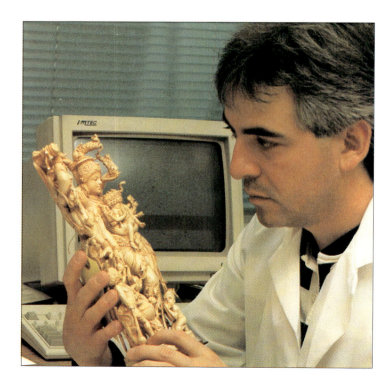

At the 1992 conference of CITES, the Convention on International Trade in Endangered Species, it was announced that the ban on trading ivory would remain in effect for at least another three years. The announcement was greeted with mixed reactions.

Some people had been afraid that ending the ban would increase poaching. They were glad the ban continued. Others were disappointed because they think the income from the legal sale of ivory could be applied to conservation efforts. They want to make wildlife preserves pay for themselves, in part by using the revenue from ivory.

Though some of the people who support the ban may agree, they don't feel that they can suddenly try to convince the buying public that it's now all right to buy ivory because it helps the elephants. For years they have been saying that buying ivory hurts the elephants' chances for survival.

The debate is likely to go on for many years along with the crucial question: What is the most economical way to manage the natural resources of Africa, and of Tanzania in particular? The answer will be vitally important to the Serengeti.

A warehouse at the U.S. National Fish and Wildlife Forensics Laboratory contains thousands of items made of wildlife parts and products from around the world. The United States has declared it illegal to buy items made from species that are endangered anywhere in the world.

Making Conservation Profitable

In a country as poor as Tanzania, residents need to get as much income from their land as possible. Any conservation method that doesn't help the people financially and that doesn't fit their normal life-style and culture won't work.

In the Serengeti, that means tourism must increase without disturbing the ecology. Tourists will continue to visit only as long as the wilderness is preserved. This is up to the leaders of the government, as well as those who elect the leaders.

Recently, the Tanzanian government implemented a new policy in some areas of the country called *sustained utilization of wildlife*. A more practical approach to wildlife conservation, this program allows regulated hunting—both for food and sport—in certain reserves. Sustained utilization of wildlife will probably become standard policy in Africa as populations continue to grow. As long as game can

SERENGETI AND MASAI TOGETHER

The Masai have been forced to live on smaller and smaller parcels of land, and they must share this land with wild animals, particularly the migrating herds. While the Masai have always tolerated wildlife, this was easier to do when they could simply move out of the animals' path. But today the Masai are losing their land. Now, villages and farms in the region keep the Masai from moving freely. The wild animals are just another nuisance, eating the grass that their cattle could eat.

If the Masai continue to lose land, their ability to graze cattle—an essential part of their lives—will disappear. Many Masai people believe that this is what must happen. Instead of raising cattle, they are turning to growing crops for a living, even though much of their land is not suitable for agriculture. Most Masai think money and education will take them into the future.

Changes are taking place. Some tourism money is now going to the Masai people, and they are also being allowed access to government-sponsored schools and medical clinics. But the Masai need more help to hang on to the grazing land that remains, so that their cattle-based culture can be preserved.

Some Masai people have gone on to higher education, attending universities in Europe and the United States. They are now returning to help their fellow Masai understand the changes that must take place in their lives. One of these individuals is Tepilit Ole Saitoti, who was featured in a National Geographic documentary, Man of the Serengeti.

"My people are in distress; they are crying out for help," said Saitoti. "They are determined to live."

be managed properly, wildlife will pay for itself and a share of the profits will go to native peoples. Tourists can continue to come and enjoy the wildlife, and local residents will have enough income to live.

The Tanzanian government is also working to bring in more tourists by improving accommodations at the lodges and hotels in the Serengeti region and by building new ones. Due to political difficulties during the 1970s and 1980s, the tourism industry that had once flourished in Tanzania's well-known national parks came to a near standstill. Services were severely cut, the lodges were neglected, roads fell into disrepair, and tourists could not even obtain such basic things as toilet paper and lightbulbs. Today, this has changed, as the government has brought tourists and their money back to the region.

Tanzanians are also promoting the country's national parks to local citizens. A government report encourages the people to "get to know and appreciate their own country and therefore feel more responsible for its preservation." There are now active public education programs in place.

These children who crowd into the city of Nairobi, Kenya, live quite near the Serengeti. The governments of Kenya and Tanzania must balance the people's need for food and jobs with the need to conserve wildlife and open land.

Encroaching on the Environment

Perhaps the major threat to the Serengeti is human encroachment, as more and more people take up residence around the park and use up its resources. This is happening in much of Africa, and Tanzania is no exception. The country's population is growing rapidly, at about 3.5 percent each year, meaning that it will probably double in the next 20 years. Towns and villages are springing up near many of the park borders, often in the game control areas. Though hunting is not allowed in many areas, other human activities are.

The people living in cities and towns need food, so farmland is developed on what was once wilderness. Cattle are left to graze on the plains. When the migrating herds pass outside the park and through these regions, it is difficult for them to find the food they need. The wildebeests and other animals will be endangered as more and more people move into the area.

Not only will the animals have a hard time, but so will the people, since there is a limit to how much food the land can provide. This is particularly true in times of drought, when no rains come to help the vegetation grow strong and healthy. Much of eastern Africa has withered in the grip of a terrible drought that has lasted through the 1980s and into the 1990s.

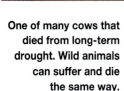

One of many cows that died from long-term drought. Wild animals can suffer and die the same way.

Again, the best solution seems to be helping the people realize that their best chance for long-term financial success is tourism and sustained utilization of wildlife. The Serengeti Ecosystem is a wilderness and wildlife reserve unlike anything else on the planet. As long as it is protected, there will always be people who will visit, contributing to the local economy.

The trick will be to make sure that both tourism and the policy of sustained utilization of wildlife are handled in ways that will both stimulate the local economy and preserve the environment. If this can be accomplished, not only will the Serengeti Plain and its animals be preserved, but they will continue to be enjoyed by people for many generations to come.

Bernhard Grzimek summed up these thoughts best when he wrote: "So much of Africa is dead already, must the rest follow? Must everything be turned into deserts, farmland, big cities, native settlements and dry brush? One small part of the continent at least should retain its original splendor so that the Africans and other peoples who follow us will be able to see it in its awe-filled past glory.

"Serengeti, at least, shall not die."

A herd of female impala on the Serengeti Plain

GLOSSARY

caldera – the bowllike depression formed when a volcano erupts and then collapses inward as it cools.

ecosystem – a group of living organisms in a particular physical environment, where the numbers and kinds of plants and animals remain balanced.

evolution – a slow, ongoing process in which a plant or animal species changes into a different form.

fossil – the physical remains or traces of a living plant or animal from prehistoric times.

habitat – the total physical and biological environment in which a plant or animal lives.

migration – the seasonal movement of animals from one area to another.

nocturnal – active primarily at night.

poaching – the illegal hunting and killing of protected animals.

predators – animals that hunt and kill other animals for food.

prey – animals, usually plant-eaters, that are killed by other animals as food.

savanna – open grassland with some trees and shrubs.

scavengers – animals that eat the meat of animals that died naturally or were killed by other animals or humans.

territory – the specific region beyond which an animal rarely travels.

ungulates – hoofed animals.

zoologist – a scientist who studies animals.

FOR MORE INFORMATION

Books

Bailey, Jill. *Save Our Species: Mission Rhino.* Austin, Tex.: Steck-Vaughn, 1990.

Bailey, Jill. *Save Our Species: Operation Elephant.* Austin, Tex.: Steck-Vaughn, 1991.

Caras, Roger. *The Endless Migration.* New York: E.P. Dutton, 1985.

Grzimek, Bernhard. *Among Animals of Africa.* New York: Stein and Day, 1970.

Grzimek, Bernhard, and Michael Grzimek. *Serengeti Shall Not Die.* New York: E. P. Dutton, 1961.

Leakey, Mary D. *Disclosing the Past.* Garden City, N.Y.: Doubleday, 1984.

Oram, Liz, and R. Robin Baker. *Mammal Migration.* Austin, Tex.: Steck-Vaughn, 1992.

Penny, Malcolm. *Rhinos—Endangered Species.* New York: Facts on File, 1988.

Peters, Lisa Westberg. *Serengeti.* New York: Crestwood House, 1989.

Sattler, Helen Roney. *Giraffes: The Sentinels of the Savannas.* New York: Lothrop, Lee & Shepard, 1989.

Schaller, George B. *Golden Shadows, Flying Hooves.* Chicago: The University of Chicago Press, 1983.

Schaller, George B. *Serengeti, A Kingdom of Predators.* New York: Alfred A. Knopf, 1972.

Schlein, Miriam. *Jane Goodall's Animal World: Hippos.* New York: Atheneum, 1989.

White, Stewart Edward. *The Land of Footprints.* Garden City, N.Y.: Doubleday, 1927.

Video

Africa's Poaching Wars. Mutual of Omaha's Spirit of Adventure, 1991.

Guide to the Greatest Wildlife Spectacles of Africa, featuring Parks of Kenya and Rwanda. Includes Masai Mara. Armchair Safaris, Ecoventures, 1988.

Guide to the Greatest Wildlife Spectacles of Africa, featuring Parks of Tanzania and Kenya. Includes Serengeti. Armchair Safaris, Ecoventures, 1988.

Life on Earth, vol. 11: "The Hunters and the Hunted." Video Classics, 1978.

Man of the Serengeti. National Geographic Society Special, 1991.

Serengeti Diary. National Geographic Society Special, 1990.

Serengeti, Part I: "No Place to Hide." Cannon Video, 1993.

Serengeti, Part II: "Drink or Die." Cannon Video, 1993.

We Live with Elephants, by Iain Douglas-Hamilton. Sierra Club Series, 1988.

INDEX